Perceptual Skills for Children

Dr. Adrianos Morakis

Kelsiehanson Verlag

DEDICATION

To My Wonderful Readers

CONTENTS

ACKNOWLEDGMENTS

Thank you to everyone who strives to grow and help others grow

1 ABSTRACT

The purpose of this study was to examine the task conditions under which external or internal attentional focus manipulations affect low or high complexity motor skills in a field study. The participants were 57 novice girls aged 14-15 years, randomly assigned to two experimental groups (the external and internal focus of attention), and a control group. The groups followed a 12 training unit intervention program to learn a simple skill (volleyball pass) and a complex volleyball skill (overhead tennis serve) and participated in a pre-test, post-test, and transfer test in the field in real sport conditions. External focus of attention group significantly improved both movement form and outcome of the simple sport skill as well as movement outcome of the complex skill. However, there was no significant difference between the different attention focus groups on the movement form of the complex skill. In conclusion, it seems that the adoption of an external focus of attention has a positive effect on both the movement form and the movement outcome of a novel simple sport skill, or the movement outcome of the more complex sport skill.

Index Terms— environmental demands; attention demands; skill complexity

2 INTRODUCTION

When novices are learning a new skill, they direct their attention to various aspects of the movement pattern, or the movement outcome (Wulfet al., 2000). Instructors direct learners' attention to the most useful information of their movement form or outcome (Beilock et al., 2004; Ford et al., 2005; Wulf et al., 1998). The positive effects of external vs internal attentional focus for motor skills learning have been proposed by many researchers (for reviews see Lohse et al., 2012; Marchant et al., 2011; Peh et al., 2011; Wulf, 2007, 2013; Wulf&Lewthwaite 2016). External focus of attention effectively speeds up the learning process so that a higher skill level is achieved sooner (Wulf, 2007), since movement kinematics resemble those typically seen at later stages of learning (Wulf&Lewthwaite, 2016). The negative effects of internal attentional focus have been explained by the "Constrained Action Hypothesis" (Wulf, McNevin, & Shea, 2001; Wulf, Shea, & Park, 2001), according to which internal focus induces a more conscious control that constrains the individual's automatic control processes of the motor system. External focus of attention, on the contrary, leads to a more automatic process. According to the "Optimal Theory" proposed by Wulf and Lewthwaite (2016), external focus serves as an important contributor to goal-action coupling. This theory describes that directing the focus of attention away from the body but using an external attentional focus (to the intended movement effect, or task goal), drives on the establishment of effective neural connections critical to effective performance and learning (Abdollahipour et al.,2017; Tsetseli et al., 2016). The external focus directs to an unconscious control process, directs movements with relative clarity toward the action goals enhancing goal-action coupling, and promotes automaticity (Wulf, McNevin, & Shea, 2001;Wulf, Shea, & Park, 2001).

It appears, however, that the external focus of attention is not frequently adopted in applied sport settings. Two issues that may contribute to this disparity are the variability of the environment (stable/changed),and the different complexity of the sports skills (low/high) or the main task (coordination task / perceptual-motor task, etc.). In sport settings, the concern is primarily learning "how" to bring about the coordinated or skilled control of simple or more often complex movements, for which the quality of movement execution is of primary importance (Shmuelof&Krakauer, 2011).

The positive influence of the external focus of attention on the learning of movement form and outcome of motor skillshas been supported by many researchers (Wulf, 2013; Lohse et al., 2012; Marchant et al., 2011) for a wide variety of skills and tasks. Wulf, McNevin, and Shea (2001), using a dynamic balance task (stabilometer), showed that the external focus group produced generally smaller balance errors than the internal focus group. Highlighting the previous results, Wulf et al. (2007) also demonstrated the superiority of anexternal focus of attention in a balance skill on surfaces of differing stability. However, there are contradictory results that indicate that the external focus of attention instructions might not be appropriate for tasks in which the main objective is to produce the correct movement form, such as the complex tasks of gymnastics routines (Lawrence et al. 2011). Additionally, Wulf et al. (2002) examined the effectiveness of the internal attentional focus (feedback for body movements) or external attentional focus (feedback for movement outcome) in the learning of complex motor skills for novice or expert participants and found that the type of feedback did not differentially affect movement quality (form). However, external-focus feedback resulted in greater accuracy (outcome) of the serves than internal-focus feedback during both practice and retention, independent of the level of expertise. It seems that more research is needed to investigate the effects of external versus the internal attentional focus on movement form.

Depending onthe level of expertise, several researchers have suggested that the external focus of attention is more beneficial for both novices and experts (Wulf et al., 1999; Wulf et al., 2002; Wulf& Su, 2007). Singh and Wulf (2020) examined whether the external focus impacts the performance of relatively inexperienced and experienced performers differently. Low-skilled and high-skilled volleyball players passed a volleyball continuously to a target. In the proximal focus condition, they were asked to concentrate on the "platform", whereas in the distal focus condition they were instructed to concentrate on the target. The high-skilled group's accuracy scores were higher in the distal relative to the proximal focus condition.

However, the low-skilled players' accuracy scores were higher in the proximal than the distal focus condition. The researchers concluded that the optimal distance of the external focus depends on the level of expertise when the skill requires a specific movement technique. An external focus on that technique seems to be more advantageous for low-skilled performers. In contrast, when the movement pattern becomes more automatic (high-skilled performers), a focus on the overall movement effect is more beneficial. It seems that the effect of both internal and external focus of attention is different according to the level of expertise. Pehet al. (2011) concluded that an external focus of attention has a different impact on different stages of learning. Becker and Smith (2013) in their study tested whether age, sex, or task complexity moderate the effect of attentional focus on motor learning. Children (boys and girls) and adults (men and women) were assigned to an internal or external attentional focus and were timed while riding a Double Pedaloeither with handles (simple task) or without handles (complex task) over a distance. The researchers found that although an external focus improved the outcome scores of the complex task, no differences were found for the simpler task. These findings suggest that attentional focus affects children and adults similarly, but task complexity moderates these effects.Most of the research to date has been conducted with expert or novice adults. However, for young adults, the goal of practice is first to learn the movement form and then improve the movement outcome of new sports skills. Very few researchers have investigated the effect of different focus of attention on young novices (Becker & Smith 2013),which is one aim of the present study.

Many researchers (Singer et al., 1994;Wulf&Weigelt, 1997) have studied the effect of the focus of attention on the performance and learning of complex motor skills and proposed the superiority of the external focus of attention. Another study by Anet al. (2013) examined the effects of attentional focus instructions on the learning of movement form and carry distance in novice golfers, demonstrating that both movement form and outcome are enhanced in complex skill learning by providing learners with relatively simple external focus instructions. Abdollahipour et al. (2015) evaluated outcome performance and movement quality in a complex gymnastics task using an internal and an external focus group as well as a control group; they found that the external focus resulted in superior movement form and outcome (greater jump height) to those of the internal or the control group. Tsetseli et al. (2018) found thatthe external focus of attention had a positive influence compared to internal focus for athletes, but only for the movement form scores, not the outcome scores. Peh et al. (2011) recommend that an important step forward in the literature may be to develop our understanding of how the different types of instruction may be directed towards movement dynamics (the form of the movement) and movement effects (the outcome of the movement) in acquiring a skill. It seems that the complexity factor might interact differently with the attentional focus factor for movement form or movement outcome, and this needs further research. Wulf and Shea (2002) suggest that studies with complex tasks under realistic conditionscould reveal new insights into feedback functions that cannot be obtained if we restrict our view to laboratory tasks (with limited freedom and sources of feedback).

The purpose of the present study was to examine the task conditions under which attentional focus manipulations have the same or different effects on low or high complexity volleyball skills with young novice learners for both movement form and outcome in a field study. Determining task characteristics that are sensitive to attentional focus manipulations will help isolate the mechanism(s) giving rise to the effect and is important for theoretical reasons. It is also important for practical reasons for teachers, coaches, or therapists to appropriately use the benefits of an external focus for motor skills of different complexity. In the present study, it was hypothesized that the external focusing of attention would improve the form and outcome of simple and complex skills for novice young participants with a transfer test in a field study.

3 MATERIALS AND METHODS

Participants

The participants were 57 novice girls aged 12-13 years (M=12.4), who were selected and randomly assigned to two experimental groups (the external or internal focus of attention), and a control group. The participants had just joined a volleyball sports club and had no previous experience in volleyball. Their participation in the research was voluntary and with the consent of the parents and coaches. Before the research, researchers and coaches agreed that during the intervention the participants would not practice further the skills to be examined (serve and passing), so as not to affect the results of the research.

Experimentaldesign

The girls participated in a set of measurements and an intervention program to learn a simple volleyball task (passing), and a more complex volleyball task (overhead tennis serve). All groups pre-tested. Then they followed an intervention program for 12 training units (6 weeks X 2 times a week) in which the two experimental groups participated. The first experimental group received explicit rules for the movement focusing attention on the body parts, while the second group received instructions to focus their attention on the target or the trajectoryof the ball. The duration of each training unit was 60 minutes: 10 minutes warm-up and 50 minutes of practical training on the field. A post-test for the acquisition phase and a transfer test followed after two weeks without training in different parameters (distance, direction). The tests and the intervention program were implemented in the field, in sport conditions.

Intervention program

All participants, except the control group, participated in the intervention program which involved practicing both motor skills (passing and serving), but with different attention focusing instructions depending on the group (external or internal). The intervention program had a total duration of 6 weeks (2 training units per week). Each training unit lasted 60 minutes: 10 minutes of warm-up on the court and 50 minutes of practicing the two skills: a) pass with fingers and b) overhead tennis serve.

Testing procedure

Each participant executed 23 frontal volleyball passes (first 3 trials) and 23 overhead tennis volleyball serves. Movement form was evaluated by subjective rating by two independent raters according to three movement quality criteria for the correct movement form: a) legs, b) torso, and c) frontal gaze for the volleyball pass, and a) body facing the direction of the serve, b) movement of the hand (archers' stance), and c) movement of the hand hitting the ball for the serve, collecting up to three (3) points. Movement outcome was evaluated according to the participants' accuracy to a target (Wulf et al., 2002), collecting up to three (3) points according to the distance from the target (3 points for 1m, 2 points for 1-2m, 1 point for 3m, 0 points for over 3m from the target). No feedback was given to the participants during the test.Before the participants started the test, the researchersdescribed the test and the evaluation methodto the participants in detail, and then a former professional volleyball player executed the task.During the transfer measurement, both the direction and the distance of the pass changed.

The volleyball pass was considered a relatively simple skill since the learners had to push the ball with both hands to a short target (2.5m distance). The volleyball serve was used as a complex skill (Wulfet al., 2002), and each participant executed overhead tennis servesto aim at a longer target (9m distance). Additionally, the volleyball serve is more difficult since it requires the participation of more body parts, more joints, more coordination with a moving object (ball), spatiotemporal coincidence, accuracy, and also a perception of the ball's trajectory, space and distance, shoulder strength and target diversity.Finally,the demands placed on the learner's memory and processing capacity arehigher than the tennis serve,which poses fewerchallenges to the learner's cognitive capacity (Wulf& Shea, 2002).

Statistics

A mixed factorial analysis of variance (3 methods X 2 complexity X 3 tests) was conducted with repeated measures on the last factor, for the evaluation of differences between the internal and external focus of attention for the different complexity skills, in three testing periods for both movement form and outcome. Whenever statistically significant differences were found, a posthoc Tukeytest was conducted.

3. Results

There was a statistically significant interaction among the 3 factors (3 methods X 2 complexity X 3 tests) ($F_{(4.54)}$ = 4.425, p <0.05): a) 3 methods X 2 complexity ($F_{(2, 54)}$ = 22.804, p <0.05), b) 3 methods X 3 tests ($F_{(4.54)}$ = 10.066, p <0.05), c) 2 complexity X 3 tests ($F_{(2.54)}$ = 14.404, p <0.05). The results are depicted in Table 1.

Table 1.Factorial ANOVA (Between-Subjects Effects) for methods, tests, and complexity, in terms of movement form and outcome.

Source	Dependent Variable	Sum of Squares	df	Mean Square	F	Sig.
method	form	11.57	2	5.78	100.21	.000
	outcome	4.68	2	2.34	86.72	.000
complexity	form	.09	1	.09	1.56	.213
	outcome	296.05	1	296.05	10971.44	.000
test	form	7.67	2	3.83	66.48	.000
	outcome	2.21	2	1.10	41.07	.000
method * complexity	form	.05	2	.029	.50	.603
	outcome	2.50	2	1.25	46.44	.000
method * test	form	3.36	4	.84	14.57	.000
	outcome	1.54	4	.38	14.27	.000
complexity * test	form	.13	2	.06	1.16	.314
	outcome	1.38	2	.69	25.59	.000
method * complexity * test	form	.04	4	.01	.19	.939
	outcome	.88	4	.22	8.15	.000

There was a statistically significant main effect of methods ($F_{(2,54)} = 21.730$, $p < 0.05$) and tests ($F_{(4,108)} = 18.638$, $p < 0.05$), as well as interaction among methods and tests ($F_{(4,108)} = 18.834$, $p < 0.05$) for the movement form of the simple skill. The results from the post hoc tests are depicted in Table 2.

Table 2.Analyses of main effects of post hoc Tukey tests among the methods and tests in the simple skill (passing) in terms of movement form.

	external (ex)		internal (in)		control (co)		
	M	*SD*	*M*	*SD*	*M*	*SD*	*Post hoc*
Pre-test (Pre)	1.68	.08	1.70	.09	1.71	.06	ex = in = co
Post-test (Post)	2.30	.07	2.09	.10	1.69	.02	ex > in > co
Transfer test (T)	2.29	.07	1.99	.04	1.68	.03	ex > in > co
Post hoc	Pre<Post, Post=T		Pre<Post, Post=T		Pre=Post, Post=T		

There was a statistically significant maineffect of methods ($F_{(2,54)} = 21.37$, $p < 0.05$), of tests ($F_{(2,108)} = 111.01$, $p < 0.05$), as well as interaction among methods and tests ($F_{(4,108)} = 21.51$, $p < 0.05$) for the movement form of the complex skill. The results from the post hoc tests are depictedin Table 3.

Table 3.Analyses of main effects of post hoc Tukey tests among the methods and tests in the complex skill (serving) in terms of movement form.

	external (ex)		internal (in)		control (co)		
	M	*SD*	*M*	*SD*	*M*	*SD*	*Post hoc*
Pre-test (Pre)	1.75	.13	1.68	.18	1.63	.24	ex = in= co
Post-test (Post)	2.37	.28	2.38	.34	1.68	.20	ex = in> co
Transfer test (T)	2.36	.29	2.37	.33	1.72	.20	ex = in> co
Post hoc	Pre<Post, Post=T		Pre<Post, Post=T		Pre=Post, Post=T		

There was a statistically significant maineffect of methods ($F_{(2,54)}$=39.54, p <0.05), of tests ($F_{(2,108)}$=99.018, p <0.05), as well as interaction among methods and tests ($F_{(4,108)}$=39.34, p <0.05) for the movement outcome of the simple skill. The results from the post hoc tests are depictedin Table 4.

Table 4.Analyses of main effects of post hoc Tukey tests among the methods and tests in the simple skill (passing) in terms of movement outcome.

	external (ex)		internal (in)		control (co)		
	M	*SD*	*M*	*SD*	*M*	*SD*	*Post hoc*
Pre-test (Pre)	2.23	.04	2.19	.02	2.21	.06	ex = in = co
Post-test (Post)	2.81	.02	2.42	.04	2.20	.03	ex > in > co
Transfer test (T)	2.80	.05	2.40	.03	2.18	.03	ex > in > co
Post hoc	Pre<Post, Post=T		Pre<Post, Post=T		Pre=Post, Post=T		

There was a statistically significant main effect of methods ($F_{(2,54)}=26.85$, $p <0.05$), of tests ($F_{(2,108)}=81.13$, $p <0.05$), as well as interaction among methods and tests ($F_{(4,108)}=34.86$, p< $p <0.05$) for the movement outcome of the complex skill. The results from the post hoc tests are depictedin Table 5.

Table 5.Analyses of main effects of post hoc Tukey tests among the methods and tests in the complex skill (serving) in terms of movement outcome.

	external (ex)		internal (in)		control (co)		
	M	*SD*	*M*	*SD*	*M*	*SD*	*Post hoc*
Pre-test (Pre)	11.31	6.83	11.31	7.78	10.26	8.57	ex = in = co
Post-test (Post)	44.47	16.57	22.10	11.21	12.36	6.74	ex > in > co
Transfer test (T)	42.89	16.77	19.21	11.57	10.00	7.63	ex > in > co
Post hoc	Pre<Post, Post=T		Pre<Post, Post=T		Pre=Post, Post=T		

Discussion

This study examined how external or internal focus of attention instruction would affect movement form and outcome of simple and more complex volleyball skills for young novices in a field study. The results showed that the external focus of attention improved both movement form and outcome of the simple skill (volleyball pass), and the movement outcome of the more complex skill (volleyball serve). However, there was no difference between the external and internal groupsin the movement form of the more complex skill.

The superiority of the external focus of attention for both movement form and outcome of simple skills found in this research is in line with Wulf's review (2013) in which she argues that external focus of attention has beneficial effects on movement effectiveness (e.g. accuracy, consistency, and balance). Moreover, movement form, analyses of movement kinematics and kinetics used in some studies have shown that whole-body coordination patterns seem to be optimized via external focus as well (An et al., 2013; Southard, 2011; Lohse et al., 2010; Wulf et al., 2010; Wulf&Dufek, 2009; Parr & Button, 2009). Instructions leading to an external focus of attention can enhance learning, contradicting the traditional view that a focus on body movement is essential for learning (Wulf et al., 2002). Our study also found that the performance benefits were not temporary. Wulf (2007), and Totsika and Wulf (2003), conclude that focusing externally leads to superior performance in delayed retention and transfer tests.

The external focus of attention group positively influenced the movement outcome of the complex skill. This result is in line with other studies (Wulf 2013) showing that external focus of attention results in more accurate outcomes than internal for volleyball serves and soccer kicks (Wulf et al., 2002), soccer throw-ins (Wulf et al., 2010), and football kicking (Zachry, 2005). Wulf and Prinz (2001) explain that the motor system is constrained when internal attentional foci interfere with the body's natural control processes. Wulf, McNevin, and Shea (2001) propose the "Constrained Action Hypothesis", explaining that an internal focus of attention brings a conscious type of control, causing individuals to constrain their motor system by interfering with automatic control processes. An external focus of attention, on the contrary, encourages a more automatic mode of control by operating unconscious, fast, and reflexive control processes. In fact, it has been proposed that using external focus instructions, the motor system can harness inherent self-organization processes to regulate movements with fewer conscious processes (Davids et al., 2008). Other studies have shown an association of external focus instructions and various measures of automaticity, including demonstrations of reduced attentional-capacity demands (Wulf, McNevin, & Shea, 2001), high-frequency movement adjustments (e.g. McNevin et al., 2003; Wulf, Shea, & Park, 2001), and reduced pre-movement times, representing more efficient motor planning (Carpenter et al., 2012).

It was also found that there was no difference between the external and internal groupson the movement form for the complex skill (volleyball serve). This finding is inconsistent with previous research that manipulated attention to external and internal features (Wulf, 2013; Wulf et al., 2002). This disparity between the current results and previous research is probably due to the different complexity level of this task. It was found that the movement form scores of both the external and internal attention focus groups showed a small improvement from pre to post and transfer test. Participants in the external focus group were asked to focus on the trajectory of the ball or the target while executing a complex skill. It is possible that asking participants to focus on trajectory or outcome while performing an attention-demanding complex skill exceeded their attentional capacity and restricted their improvement. Additionally, the lack of declarative knowledge probably decreased the movement form scores of the external focus group. Although the participants in the internal focus group were instructed to focus their attention on a single point (of their body part), they still had to execute a complex skill that demands much attention. The internal focus of attention participants received declarative knowledge of how to execute the movement form of the complex skill. However, their efforts to implement the instructions disrupted the automatic control processes of the motor system, decreased the goal-action coupling, and decreased the improvement in movement form. These findings are consistent with previous studies that explicitly assessed movement form (Lawrence et al. 2011; Wulf et al., 2002), and it appears that when the tasks are assessed on the accurate production of a particular movement pattern/technique or form, the benefits of adopting an external focus of attention cannot be generalized. Peh and colleagues (2011) assert that the internal focus of attention may also be useful to learners if the performance context emphasizes movement form instead of performance outcomes. It seems that the external focus of attention is not frequently adopted in applied sport settings when the goal is to improve movement form. Similarly, Lawrence et al. (2011) propose that the external focus of instructions might not be a priority for tasks in which the main objective is to produce the correct movement form. Two studies by Maddox et al. (1999) and Wulf, McNevin, & Shea (2001) have used expert ratings to evaluate movement quality under external versus internal attentional focus conditions; both of these studies reported no differences in movement form.

As Wulf and colleagues (2002) remark, "in the acquisition of complex motor skills with many degrees of freedom, such as the ones used here, where the performer's major goal is the learning of the correct technique, it might be almost impossible to come up with statements that do not refer at all to the performer's own movements" (p. 176). Lawrence and colleagues (2011) argue that "the benefits of adopting an external focus do not generalize to tasks that are assessed on the accurate production of a particular movement pattern/technique or form" (p.439).Perhaps focusing on the movement technique, while not concentrating on the movements per se, might be more beneficial than focusing on the outcome directly. This may be especially true for novices attempting to acquire a relatively complex skill such as the one used here (Wulf et al., 2000). Peh and colleagues (2011) mention that "an internal focus of attention may be more useful to the learner if the performance context emphasizes movement form instead of performance outcomes" (p. 73), and go on to say that further research is neededon tasks which emphasize the production of specific movement forms. Similarly, Lawrence et al. (2011) propose that external focus instructions might not be appropriate for tasks in which the main objective is to produce the correct movement form, such as gymnastics routines.

In conclusion, it seems that the adoption of an external focus of attention (directing the attention to the target or the trajectory of the ball) has a positive effect on both the movement form and the movement outcome of a novel simple sport skill, or the movement outcome of the more complex sport skill. However, when learning a more complex novel sport skill, the adoption of either an internal (directing attention to body parts) or an external focus of attention instruction method improves movement form. Due to the ambiguity of the results on the positive effects of external foci on the movement form of complex skills, more research is needed to identify the positive effects of external focus on movement form for sports skills. Additionally, due to the different complexity of sports skills even in the same sport, the task constraints, the different sports environment demands, the different evaluation criteria (movement form or outcome), and the practice goals in different stages of learning, more studies are needed to reevaluate whether the external focus of attention instructions are beneficial in these different sport situations.

References

[1] Abdollahipour, R., Nieto, M. P., Psotta, R., &Wulf, G. (2017). External focus of attention and autonomy support have additive benefits for motor performance in children. *Psychology of Sport and Exercise*, 32, 17-24.

[2] Abdollahipour, R., Wulf, G., Psotta, R., &Palomo Nieto, M. (2015). Performance of gymnastics skill benefits from an external focus of attention. *Journal of Sports Sciences, 33*(17), 1807-1813.

[3] An, J., Wulf, G., & Kim, S. (2013). Increased carry distance and X-factor stretch in golf through an external focus of attention. *Journal of Motor Learning and Development, 1*(1), 2-11.

[4] Becker, K., & Smith, P. J. (2013). Age, task complexity, and sex as potential moderators of attentional focus effects. *Perceptual and Motor Skills, 117*(1), 130-144.

[5] Beilock, S. L., Bertenthal, B. I., McCoy, A. M., & Carr, T. H. (2004). Haste does not always make waste: Expertise, direction of attention, and speed versus accuracy in performing sensorimotor skills. *PsychonomicBulletin &Review, 11*(2), 373-379.

[6] Carpenter, S. K., Lohse, K. R., Healy, A. F., BourneJr, L. E., & Clegg, B. A. (2012). External focus of attention improves performance in a speeded aiming task. *Journal of Applied Research in Memory and Cognition, 2*(1), 14-19.

[7] Davids, K. W., Button, C., & Bennett, S. J. (2008). *Dynamics of skill acquisition: A constraints-led approach.* Human Kinetics.

[8] Ford, P., Hodges, N. J., & Williams, A. M. (2005). Online attentional-focus manipulations in a soccer-dribbling task: Implications for the proceduralization of motor skills. *Journal of Motor Behavior, 37*(5), 386-394.

[9] Lawrence, G. P., Gottwald, V. M., Hardy, J., & Khan, M. A. (2011). Internal and external focus of attention in a novice form sport. *Research Quarterly for Exercise and Sport, 82*(3), 431-441.

[10] Lohse, K. R., Sherwood, D. E., & Healy, A. F. (2010). How changing the focus of attention affects performance, kinematics, and electromyography in dart throwing. *Human Movement Science, 29*(4), 542-555.

[11] Lohse, K. R., Wulf, G., &Lewthwaite, R. (2012).Attentional focus affects movement efficiency. In N. Hodges &A. Mark Williams (Eds.), *Skill acquisition in sport: Research, theory & practice*(pp. 40-58).Routledge.

[12] Maddox, M. D., Wulf, G., & Wright, D. L. (1999).The effect of an internal vs. external focus of attention on the learning of a tennis stroke. *Journal of Exercise Psychology, 21*(2), 78-85.

[13] Marchant, D. C., Greig, M., Bullough, J., &Hitchen, D. (2011). Instructions to adopt an external focus enhance muscular endurance. *Research Quarterly for Exercise and Sport*, 82(3), 466-473.

[14] Parr, R., & Button, C. (2009).End-point focus of attention: Learning the" catch" in rowing. *International Journal of Sport Psychology, 40*(4), 616.

[15] Peh, S. Y. C., Chow, J. Y., &Davids, K. (2011). Focus of attention and its impact on movement behaviour. *Journal of Science and Medicine in Sport, 14*(1), 70-78.

[16] Shmuelof, L., &Krakauer, J. W. (2011). Are we ready for a natural history of motor learning?. *Neuron, 72*(3), 469-476.

[17] Singer, R. N., Lidor, R., &Cauraugh, J. H. (1994). Focus of attention during motor skill performance. *Journal of Sports Sciences, 12*(4), 335-340.

[18] Singh, H., &Wulf, G. (2020). The distance effect and level of expertise: Is the optimal external focus different for low-skilled and high-skilled performers?. *Human Movement Science*, *73*, 102663.
[19] Southard, D. (2011). Attentional focus and control parameter: Effect on throwing pattern and performance. *Research Quarterly for Exercise and Sport*, *82*(4), 652-666.
[20] Totsika, V., &Wulf, G. (2003).The influence of external and internal foci of attention on transfer to novel situations and skills. *Research Quarterly for Exercise and Sport*, *74*(2), 220-232.
[21] Tsetseli, M., Zetou, E., Vernadakis, N., &Michalopoulou, M. (2016).The effect of internal and external focus of attention on game performance in tennis. *ActaGymnica*, *46*(4), 162-173.
[22] Tsetseli, M., Zetou, E., Vernadakis, N., &Mountaki, F. (2018). The attentional focus impact on tennis skills' technique in 10 and under years old players: Implications for real game situations.
[23] Wulf, G. (2007). *Attention and Motor Skill Learning*. Human Kinetics.
[24] Wulf, G. (2013). Attentional focus and motor learning: a review of 15 years. *International Review of Sport and Exercise Psychology*, *6*(1), 77-104.
[25] Wulf, G., &Dufek, J. S. (2009). Increased jump height with an external focus due to enhanced lower extremity joint kinetics. *Journal of Motor Behavior*, *41*(5), 401-409.
[26] Wulf, G., &Lewthwaite, R. (2016).Optimizing performance through intrinsic motivation and attention for learning: The OPTIMAL theory of motor learning. *PsychonomicBulletin &Review*, *23*(5), 1382-1414.
[27] Wulf, G., McNevin, N., & Shea, C. H. (2001).The automaticity of complex motor skill learning as a function of attentional focus. *The Quarterly Journal of Experimental Psychology Section A*, *54*(4), 1143-1154.
[28] Wulf, G., &Prinz, W. (2001). Directing attention to movement effects enhances learning: A review. *PsychonomicBulletin &Review*, *8*(4), 648-660.
[29] Wulf, G., & Shea, C. H. (2002). Principles derived from the study of simple skills do not generalize to complex skill learning. *PsychonomicBulletin &Review*, *9*(2), 185-211.
[30] Wulf, G., & Su, J. (2007). An external focus of attention enhances golf shot accuracy in beginners and experts. *Research Quarterly for Exercise and Sport*, *78*(4), 384-389.
[31] Wulf, G., &Weigelt, C. (1997). Instructions about physical principles in learning a complex motor skill: To tell or not to tell…. *Research Quarterly for Exercise and Sport*, *68*(4), 362-367.
[32] Wulf, G., Chiviacowsky, S., Schiller, E., &Ávila, L. T. G. (2010). *Frequent ExternalFocus.*
[33] Wulf, G., Höß, M., &Prinz, W. (1998). Instructions for motor learning: Differential effects of internal versus external focus of attention. *Journal of Motor Behavior*, 30(2), 169-179.
[34] Wulf, G., McConnel, N., Gärtner, M., & Schwarz, A. (2002).Enhancing the learning of sport skills through external-focus feedback. *Journal of Motor Behavior*, *34*(2), 171-182.
[35] Wulf, G., McNevin, N. H., Fuchs, T., Ritter, F., & Toole, T. (2000).Attentional focus in complex skill learning. *Research Quarterly for Exercise and Sport*, *71*(3), 229-239.
[36] Wulf, G., Shea, C., & Park, J. H. (2001). Attention and motor performance: preferences for and advantages of an external focus. *Research Quarterly for Exercise and Sport*, *72*(4), 335-344.
[37] Wulf, G., Töllner, T., & Shea, C. H. (2007).Attentional focus effects as a function of task difficulty. *Research Quarterly for Exercise and Sport*, *78*(3), 257-264.
[38] Zachry, T. L. (2005). *Effects of attentional focus on kinematics and muscle activation patterns as a function of expertise*.[Master's thesis, University of Nevada, Las Vegas].UNLV Retrospective Theses & Dissertations. 1862.

Made in United States
Orlando, FL
04 November 2026

90206049R00015